Erica Glessing

MASTER YOUR BUSINESS LISTING

Attract Clients Easily & Get Seen Online

Paperback ISBN: 978-1-957142-27-2

Parts of this book were transcribed from a Google business profile workshop held by author Erica Glessing. AI was utilized in the creation of this book.

Google Business Listing Explained: A Clearer Picture

It might seem too easy. The google business listing is a phenomenal tool for attracting clients to your business. Often people skip over the creation of a business listing – after all, a fantastic website could be a higher priority. Websites are great tools, but a strong business listing can attract the right clients with ease – and you don't have to pay to keep it online! In this book we'll explore what makes a listing great and how you can maximize the potential of your business listing to attract clients.

Many people are unfamiliar with Google Business Listings. Let's explore what they are and how they benefit your business.

Using a Local Example

I'll use our online visibility as an example. Let's say you search for "title company Morgan Hill" on Google Maps. This result is a Google Business Listing.

Websites vs. Business Listings

The ideal scenario for a small business would be for the business profile to show up in the "map pack" of business listings and atop the organic listings for the website. We often compare websites to business cards. They showcase your business to those who already know you. The combination of a wonderful website coupled with a vibrant Google Business Listing are key for attracting new customers searching for your services.

For instance, if someone types "family law attorney Morgan Hill," your listing should ideally appear in the map pack. This is where potential clients can find you. Larger marketing firms dominate the top organic listings on search engines. The map pack business listing is one way to shine even if you don't have the SEO budget of a massive marketing firm.

The Rise of Local SEO

In the past, search engine optimization (SEO) heavily influenced search results. Companies with extensive keyword-stuffed content ranked higher. This changed around the year 2000. Google prioritized user experience, ensuring searches yielded relevant local results.

Now, when you search for "pizza near me," Google presents nearby options, not pizzerias across the country. This is where Google Business Listings come in.

A Level Playing Field for Local Businesses

Google Business Listings help local businesses compete with larger companies that might have better traditional SEO. It prioritizes businesses relevant to a user's location. Imagine needing a print shop – you wouldn't want to fly to another state to pick up your order!

Maintaining Your Google Business Listing

Once you create a Google Business Listing, Google wants it to thrive. They encourage optimizing your listing with elements like:

» Photos: The average business has around 11 photos on their listing.

» Website Clicks: While phone calls might seem ideal, most users prefer clicking the website to learn more before contacting you.

» Positive Reviews: Reviews build trust and encourage potential customers to choose you.

Embrace the Power of Google Maps

By creating and maintaining a Google Business Listing, you increase your local visibility and attract new clients who are actively searching for your services in your area.

Table of Contents

Chapter 1

Understanding the Importance of Google Business Listings: A Guide to Local SEO Domination

In today's digital age, a strong online presence is no longer a luxury for businesses – it's a necessity. For local businesses, particularly, Google Business Listings (GBL), now known as Google Business Profile (GBP), play a crucial role in attracting customers and driving growth. This guide dives deep into understanding the importance of GBP, exploring its benefits, optimization strategies, and best practices to leverage its power for local SEO dominance.

Why Google Business Profile Matters

Imagine a potential customer searching for a bakery near them. Google, recognizing their local intent, displays a map with relevant listings at the top of the search results page – this is the coveted "local pack." A well-optimized GBP increases your chances of appearing in this prime location, significantly boosting your visibility. Here's a breakdown of the key benefits GBP offers:

» Increased Visibility: GBP is a free tool that allows you to manage how your business appears on Google Search and Maps. By claiming and optimizing your profile, you ensure your business information, including address, phone number, website, and photos, is accurate and up-to-date. This increases the likelihood of your business appearing in relevant local searches, bringing you closer to potential customers who are actively looking for what you offer.

» Enhanced Credibility: GBP allows you to showcase customer reviews, ratings, and Q&A sections. Positive reviews build trust and credibility, influencing customer decisions. Responding to reviews, both positive and negative, demonstrates your commitment to customer satisfaction and fosters a sense of transparency.

» Direct Communication Channel: Your GBP acts as a direct communication channel with potential customers. You can display your contact information, website link, and even a booking button, allowing users to connect with you seamlessly. Additionally, features like Google Posts enable you to share updates, events, and promotions directly through your profile, keeping your audience engaged.

» Free and Valuable Insights: GBP provides valuable analytics about user engagement with your listing. You can track views, website clicks, and even calls received directly from your profile. These insights help you understand how customers find you online and refine your optimization strategy for better results.

» Improved Local SEO: A well-maintained GBP is a significant factor in Google's local search ranking algorithms. By optimizing your profile with relevant keywords, categories, and high-quality photos, you improve your local SEO ranking, increasing your chances of appearing in top search results for local searches.

Optimizing Your Google Business Profile for Success

Creating a GBP is just the first step. To unlock its full potential, optimization is crucial. Here are some key strategies:

» Complete and Accurate Information: Ensure all your business details in the profile are accurate and up-to-date, including name, address, phone number, website URL, and business hours. Inconsistencies can negatively impact your ranking and confuse potential customers.

» Choose the Right Categories: Select relevant categories that accurately describe your business offerings. Google allows you to choose primary and secondary categories, so leverage this to ensure your business gets discovered by the right audience.

» Compelling Description: Craft a concise yet informative description of your business that highlights your unique selling points and value proposition. Use relevant keywords naturally to improve search ranking.

» Encourage Customer Reviews: Positive reviews are gold. Encourage satisfied customers to leave reviews on your GBP. Respond to all reviews, thanking customers for positive feedback and addressing concerns raised in negative reviews professionally.

» Regular Engagement: Don't let your profile become static. Utilize Google Posts to share updates, events, promotions, and special offers. Regularly posting fresh content keeps your audience engaged and informed about your business.

» Track and Analyze Performance: GBP offers valuable insights into user interaction with your profile. Track views, clicks, call metrics, and analyze which posts resonate with your audience. Use these insights to refine your strategy and optimize your profile for better results.

Best Practices for Your Google Business Profile

- Claim and Verify Your Listing: The first step is to claim your existing GBP or create a new one if it doesn't exist. Google requires verification to ensure legitimacy.

- Manage Multiple Locations: If you have multiple locations, create separate GBPs for each one. This ensures each location has a dedicated online presence and improves visibility in local searches.

- Monitor for Duplicate Listings: Sometimes, duplicate listings can appear. Regularly check for duplicates and claim or merge them into your primary GBP to avoid confusion for potential customers.

- Stay Updated on Google's Guidelines: Google's algorithms and guidelines are constantly evolving. Stay updated.

Chapter 2

Crafting the Perfect Shop Window: A Guide to Creating an Optimized Google Business Profile

In the bustling online marketplace, a well-designed storefront is essential for attracting customers. For local businesses, Google Business Profile (GBP) serves as your prime digital storefront, influencing how potential customers perceive and interact with your brand. This comprehensive guide equips you with the knowledge and strategies to create an optimized GBP, propelling your local SEO and attracting a steady stream of customers.

Claiming Your Digital Space: The Foundation of Optimization

The journey begins with claiming your existing GBP or creating a new one if it doesn't exist. Here's a step-by-step breakdown:

1. Navigate to google.com/business and sign in with a Google account.
2. Search for your business name. If it exists, claim it by following the on-screen prompts. If not, click "Create a Business with this Name."

3. Enter your complete and accurate business information, including name, address, phone number, and website URL. Double-check for accuracy, as inconsistencies can negatively impact your ranking and confuse potential customers.

4. Choose the primary business category that best describes your offerings. You can also select a secondary category for further detail.

Verification: The Key to Legitimacy

Once you've claimed or created your profile, Google requires verification to ensure its legitimacy. Verification methods vary depending on your business type, but common options include postcard verification, phone verification, or email verification. Prompt verification ensures your GBP appears in search results and builds trust with potential customers.

Crafting a Compelling Description: The Power of Words

Think of your GBP description as your digital elevator pitch. Here are some tips for crafting a compelling description:

» Keep it concise and informative: Aim for 750 characters or less.

» Highlight your unique selling points: What sets your business apart from the competition?

» Use relevant keywords naturally: Incorporate keywords that potential customers might use to search for your type of business.

- » Focus on the benefits: Don't just list features; explain how your products or services solve customer problems.
- » Call to action: Encourage users to visit your website, contact you, or learn more about a special offer.

Visual Storytelling: The Power of Images

In the digital age, visuals are powerful tools for grabbing attention and conveying information. Here's how to leverage images in your GBP:

- » High-quality photos are essential: Invest in professional photos or take high-quality pictures yourself.
- » Showcase your storefront, products, services, and team: Give potential customers a glimpse into your business.
- » Maintain a consistent visual style: This creates a professional and cohesive look for your profile.
- » Regularly update your photos: Keep your profile fresh and showcase new products or seasonal offerings.
- » Utilize the logo upload feature: Your logo is a visual representation of your brand; use a high-resolution version. Here's the author's business logo:

SEO
for LEAD GEN

The Power of Reviews: Building Trust and Credibility

Positive customer reviews are gold for your GBP. Here's how to encourage reviews and leverage their power:

» Request reviews politely: Ask satisfied customers to leave a review on your GBP. Offer incentives like discounts or loyalty points (within Google's guidelines).

» Respond to all reviews: Thank customers for positive feedback and address concerns raised in negative reviews professionally. This demonstrates your commitment to customer satisfaction.

» Showcase positive reviews: Strategically display positive reviews on your website or marketing materials to build trust with potential customers.

Keeping Your Customers Engaged: The Power of Google Posts

Don't let your GBP become a static online brochure. Google Posts allows you to share updates, events, promotions, and special offers directly through your profile. Here are some tips for creating engaging Google Posts:

» Post consistently: Aim for at least weekly updates to keep your audience engaged.

» Variety is key: Mix up your content with text posts, images, and videos.

» Highlight special offers and promotions: Promote upcoming sales, discounts, or seasonal deals.

» Announce events and news: Keep your audience informed about workshops, product launches, or community events.

- Use a call to action: Encourage users to visit your website, learn more about a promotion, or contact you for inquiries.

Chapter 3

Harnessing the Power of Location: How to Leverage Google Maps for Visibility

In today's digital landscape, where convenience reigns supreme, potential customers increasingly rely on online maps to find businesses nearby. Google Maps, with its dominant market share, presents a golden opportunity for businesses to enhance their visibility and attract a steady stream of customers. This comprehensive guide delves into how to leverage Google Maps for maximum visibility, turning it into a powerful marketing tool.

Claiming Your Territory: The Foundation of Map Domination

The cornerstone of leveraging Google Maps lies in claiming and optimizing your Google Business Profile (GBP), formerly known as Google My Business (GMB). Here's why a well-maintained GBP is crucial:

» Increased Visibility: A claimed and optimized GBP significantly increases your chances of appearing in relevant local searches on Google Maps. This prime placement brings you closer to potential customers actively seeking businesses like yours.

- Enhanced Credibility: Your GBP acts as your digital storefront on Google Maps. Positive reviews, high ratings, and informative content build trust and credibility, influencing customer decisions.

- Direct Communication Channel: GBP allows you to showcase your contact information, website link, and even a booking button, enabling seamless communication with potential customers.

Optimizing Your Google Business Profile: A Recipe for Success

A claimed GBP is just the beginning. To truly leverage Google Maps, optimization is key. Here are some effective strategies:

- Accurate and Complete Information: Ensure all details in your profile are accurate and up-to-date, including business name, address, phone number, website URL, and opening hours. Inconsistencies can negatively impact your ranking and confuse potential customers. This is called the "NAP" or name, address, phone. In a perfect world, all of your NAPs are in alignment from your google listing to your social media profiles and your business website. Google likes it when everything agrees.

- Category Selection: Choose relevant categories that precisely describe your business offerings. Google allows primary and secondary categories, so leverage this to ensure your business gets discovered by the right audience searching for what you offer.

- Compelling Description: Craft a concise yet informative description that highlights your unique selling points and value proposition. Use relevant keywords naturally to improve search ranking on Google Maps.
- Visual Storytelling with High-Quality Photos: Visuals are captivating. Include high-quality photos showcasing your storefront, products, services, and team. Regularly update your photos to keep your profile fresh and engaging.
- Harnessing the Power of Reviews: Positive reviews are like gold for your GBP. Encourage satisfied customers to leave reviews and respond to all reviews, both positive and negative, demonstrating your commitment to customer satisfaction.
- Regular Engagement with Google Posts: Don't let your profile become static. Utilize Google Posts to share updates, events, promotions, and special offers directly through your GBP. Regularly posting fresh content keeps your audience engaged and informed.

Beyond Optimization: Expanding Your Reach on Google Maps

While a well-optimized GBP is essential, there are additional strategies to maximize your visibility:

- Responding to Customer Questions: The Q&A section in your GBP allows users to ask questions about your business. Promptly and professionally answering these questions demonstrates transparency and excellent customer service.

- Keeping Up-to-Date with Google My Business Updates: Google's algorithms and guidelines are constantly evolving. Stay updated on the latest changes to ensure your GBP adheres to best practices and maintains optimal visibility.
- Managing Multiple Locations: If you have multiple locations, create separate GBPs for each one. This ensures each location has a dedicated presence on Google Maps, improving local search visibility for all your branches.
- Encouraging Check-Ins: Motivate customers to check in at your location on Google Maps. This social proof can influence others searching for similar businesses nearby.
- Leveraging Google Maps Ads: Consider Google Maps Ads to target specific demographics or locations with tailored promotional messages. This targeted approach allows you to reach a highly relevant audience actively searching for businesses like yours.

Tracking and Analyzing Performance: Data-Driven Decisions

GBP analytics offer valuable insights into user interaction with your profile on Google Maps. Here are some key metrics to track and analyze:

- Profile Views: This indicates how many times your business profile has been viewed in search results on Google Maps.
- Search Queries: Analyze the search terms users employed to find your business. This helps you understand

how customers discover you and allows you to refine your GBP description and keywords accordingly.

» Website Clicks: Track how often users click through to your website from your Google Maps profile. This helps you gauge the effectiveness of your GBP in driving website traffic.

» Direction Requests: Monitor the number of times users request directions to your location on Google Maps. This indicates potential customer intent and can help you evaluate the effectiveness of your location information.

A Roadmap to Success

By claiming, optimizing, and actively managing your Google Business Profile, you unlock the immense potential of Google Maps for local business visibility.

Chapter 4

Mastering Your Neighborhood: Optimizing for Local SEO

In today's digital age, where "near me" searches dominate local customer journeys, local SEO has become a critical factor for brick-and-mortar businesses and service providers with a physical presence. By optimizing your online presence for local search, you increase your visibility in search results for geographically targeted keywords, attracting customers actively seeking what you offer. This guide delves into the two core pillars of local SEO: keyword research for local search terms and implementing local SEO best practices, empowering you to dominate your local market.

Keyword Research for Local Search Terms: Unearthing Your Customers' Voices

Keyword research is the foundation of any successful SEO strategy, and local SEO is no exception. However, local search terms differ from traditional SEO keywords. Here's how to identify the specific keywords your target audience uses to find businesses like yours:

» Understanding Local Intent: Traditional SEO focuses on broad keywords, while local SEO prioritizes keywords with geographic intent. Think "best bakery San Francisco" instead of just "bakery."

» Leveraging Location-Specific Keywords: Integrate your city, town, or neighborhood into your keyword research. Tools like Google Keyword Planner allow you to filter results based on location.

» Considering Long-Tail Keywords: Long-tail keywords are more specific and often have lower competition. Examples include "dog grooming services near Los Angeles" or "hair stylist for curly hair Brooklyn."

» Utilizing Local Search Tools: Several online tools specialize in local keyword research. Explore options like Moz Local, Yelp Ads Keyword Tool, and SpyFu to discover relevant local search terms.

» Competitive Analysis: Analyze your competitor's online presence. See what keywords they rank for and identify potential gaps you can exploit with your own local SEO strategy.

» Leveraging Customer Insights: Pay attention to how your existing customers find you. What search terms do they mention using?

Implementing Local SEO Best Practices: Building a Strong Local Foundation

Once you have a solid understanding of relevant local search terms, it's time to implement best practices to optimize your online presence for local success. Here are some key strategies:

- Claim and Optimize Your Google Business Profile (GBP): A well-maintained GBP is the cornerstone of local SEO. Claim your profile, ensure all information is accurate and up-to-date, choose relevant categories, and craft a compelling description using local keywords.

- NAP Consistency: NAP stands for Name, Address, and Phone number. Ensure your NAP information is consistent across all online directories, listings, and your website. Inconsistencies can negatively impact your local SEO ranking.

- Local Citations: Citations are online mentions of your business name, address, and phone number on websites other than your own. Encourage citations on local directories, industry-specific websites, and business listings relevant to your area.

- On-Page Optimization: Optimize your website content for local search terms. Include relevant keywords naturally throughout your website, including page titles, meta descriptions, headers, and body content.

- Optimize for Mobile: A significant portion of local searches occur on mobile devices. Ensure your website is mobile-friendly for a seamless user experience.

- Positive Reviews and Customer Engagement: Positive online reviews are powerful trust signals for local customers. Encourage satisfied customers to leave reviews on your GBP and other platforms. Respond to all reviews, both positive and negative, demonstrating your commitment to customer satisfaction.

» High-Quality Local Content: Create valuable content that resonates with your local audience. This could include blog posts highlighting local events, guides to local attractions, or showcasing your expertise in your industry with a local angle.

» Engage with Local Online Communities: Participate in relevant online communities like Facebook groups or neighborhood forums. Offer helpful advice, answer questions, and establish yourself as a local expert.

» Local Link Building: Building backlinks from high-authority local websites can significantly improve your local SEO ranking. Partner with local businesses for guest blogging opportunities or sponsor local events to earn backlinks.

Tracking and Analyzing Local SEO Performance: Measuring Your Success

Local SEO is an ongoing process that requires regular monitoring and adjustments. Here are some key metrics to track and analyze your local SEO performance:

» Local Keyword Ranking: Monitor your ranking for your chosen local keywords in search results. Tools like Google Search Console can provide valuable insights.

» Website Traffic: Track website traffic coming from local searches. This helps you gauge the effectiveness of your local SEO efforts in attracting potential customers.

» Google Business Listing Insights: GBP analytics provide

valuable data on user interaction with your profile. Track profile views, website clicks, and call metrics to understand how users find you on Google Maps.

» Customer Conversion Rates: Analyze how many website visitors from local searches convert into paying customers. This helps you evaluate the overall success of your local SEO

Chapter 5

The Power of Reputation: Maximizing Reviews and Ratings in the Digital Age

In today's digital landscape, online reviews and ratings hold immense power. They act as digital word-of-mouth, influencing customer decisions and shaping brand perception. For businesses, maximizing positive reviews and effectively addressing negative ones is crucial for building trust, attracting new customers, and achieving long-term success. This comprehensive guide equips you with strategies to cultivate a thriving online reputation, turning customer feedback into a powerful marketing tool.

Encouraging Positive Reviews from Customers: Building a Stream of Advocacy

Positive reviews are gold. Here's how to encourage a steady flow of positive feedback:

» Deliver Exceptional Customer Service: The foundation of positive reviews lies in consistently exceeding customer expectations. Provide attentive, helpful, and friendly service at every touchpoint.

» Go the Extra Mile: Surprise and delight your customers with unexpected gestures of goodwill. This could be a complimentary upgrade, a handwritten thank you note, or a personalized offer.

» Make it Easy to Leave Reviews: Don't make customers jump through hoops. Provide clear and easy ways for them to leave reviews on your Google Business Profile (GBP), social media platforms like Yelp and Facebook, and industry-specific review sites relevant to your business.

» Directly Request Reviews: After a positive customer interaction, politely ask them to leave a review on your preferred platform. Offer guidance on where to find your profile and highlight the value of their feedback.

» Incentivize Reviews (Within Guidelines): Consider offering small incentives like discounts or loyalty points for leaving a review (while adhering to platform and legal guidelines).

» Respond Promptly to All Reviews: Show your customers you value their feedback by responding to all reviews, both positive and negative. Thank customers for positive reviews and acknowledge their feedback.

» Showcase Positive Reviews: Strategically display positive reviews on your website, marketing materials, and social media profiles. This builds trust and credibility with potential customers.

Turning Satisfied Customers into Vocal Advocates:

» Loyalty Programs: Reward loyal customers with a well-designed loyalty program that incentivizes repeat business and encourages positive word-of-mouth.

» Social Media Engagement: Foster a positive social media presence where customers feel comfortable leaving feedback and engaging with your brand. Respond to comments and questions promptly.

» Customer Feedback Surveys: Regularly conduct customer feedback surveys to understand customer satisfaction and identify areas for improvement. Positive feedback in surveys can be repurposed as testimonials on your website.

» Run Review Contests: Occasionally host contests or giveaways where participation requires leaving a review. This can generate a surge of new reviews in a short period.

Dealing with Negative Reviews Effectively: Turning Challenges into Opportunities

Negative reviews happen, and it's how you handle them that truly matters. Here are strategies for addressing negative feedback effectively:

» Respond Promptly: Don't ignore negative reviews. Address them quickly and professionally, ideally within 24 hours.

» Acknowledge the Customer's Frustration: Start your response by acknowledging the customer's frustration and expressing empathy for their experience.

- Offer a Sincere Apology: Apologize for the inconvenience caused, even if you believe their perception is inaccurate.
- Take Ownership of the Issue: Demonstrate responsibility for resolving the situation. Offer solutions and outline the steps you'll take to prevent similar issues in the future.
- Move the Conversation Offline: For complex situations, suggest taking the conversation offline to a private channel like email or phone to discuss specific details and reach a resolution.
- Maintain a Professional Tone: Even if the customer's language is aggressive or hostile, maintain a professional and courteous tone in your response. Avoid getting defensive or argumentative.
- Use Negative Reviews for Improvement: Negative reviews, while unpleasant, offer valuable insights into areas needing improvement. Analyze the feedback and use it to refine your processes and customer service strategies.

Building a Culture of Customer Centricity:

- Empower Your Team: Train your staff to prioritize customer satisfaction and empower them to resolve customer issues effectively.
- Regularly Solicit Feedback: Develop a system for routinely soliciting customer feedback, both positive and

negative. This can be done through surveys, social media engagement, or post-interaction emails.

» Recognize and Reward Positive Customer Interactions: Acknowledge and celebrate employees who consistently deliver exceptional customer service and generate positive feedback.

» Make Customer Satisfaction a Core Value: Embed customer centricity into your company culture. Ensure all employees understand the importance of positive reviews and their role in building a strong online reputation.

The Power of Positive Feedback

By implementing these strategies, you can cultivate a thriving online reputation, fostering trust and attracting new customers. Remember, positive reviews.

Chapter 6

Mastering Your Digital Storefront: Utilizing Advanced Google My Business Features

In today's digital landscape, Google Business Profile (GBP), formerly known as Google My Business, serves as a critical tool for local businesses. Beyond the fundamental features of claiming your profile and managing basic information, GBP offers a treasure trove of advanced features and tools designed to enhance your online presence and drive business growth. This guide delves into these advanced functionalities, empowering you to unlock the full potential of your GBP and leverage valuable insights for strategic decision making.

Showcasing Your Business with Rich Content

» High-Quality Photos and Videos: Visuals are captivating. Utilize high-quality photos showcasing your storefront, products, services, team, and satisfied customers. Consider adding a virtual tour using Google Street View for an immersive experience.

- Google Posts: Don't let your profile become static. Regularly utilize Google Posts to share updates, events, promotions, special offers, and new products or services directly through your GBP. Posts can include text, images, and even short videos.
- Q&A Section: The Q&A section allows users to ask questions about your business. Promptly and professionally answering these questions demonstrates transparency and excellent customer service. Consider pre-populating the Q&A section with frequently asked questions to anticipate customer inquiries.

Engaging Your Audience and Building Relationships

- Responding to Reviews: Actively respond to all reviews, positive and negative. Thank customers for positive feedback and acknowledge concerns raised in negative reviews. Demonstrate your commitment to customer satisfaction and a willingness to address issues.
- Messaging Functionality: Enable the messaging functionality in your GBP to allow potential customers to message you directly through your profile. This allows for real-time communication and can help answer questions or address concerns promptly.
- Short Name Customization: Claim a short name for your GBP. This creates a unique and memorable identifier that customers can easily use in online searches and navigation.
- Showcasing Attributes: Highlight unique features and

amenities offered by your business by selecting relevant attributes through GBP. This could include wheelchair accessibility, Wi-Fi availability, or kid-friendliness.

Harnessing Insights for Strategic Decision Making

- » GBP Insights: GBP offers valuable data and analytics about user interaction with your profile. Track metrics like profile views, website clicks, call metrics, and how users find your listing (discovery source).
- » Understanding Your Audience: Analyze demographics of users who view your profile. This can help you tailor your content strategy and marketing efforts to resonate with your target audience.
- » Measuring the Impact of Google Posts: Track the performance of your Google Posts. See which posts generate the most engagement and use this data to refine your content strategy for future posts.
- » Optimizing Based on Insights: Utilize the insights gleaned from GBP to inform your overall business strategy. Identify areas where you can improve your online presence or refine your offerings to better cater to customer needs.

Advanced Features for Enhanced Visibility

- » Showcase Your Products: Showcase your products directly through your GBP. Upload high-quality photos, add detailed descriptions, and highlight key features. This allows potential customers to browse your offerings directly on your profile.

» Promotions: Run targeted promotions directly through your GBP. Announce special offers, discounts, or seasonal deals to attract new customers and encourage repeat business.

» Service Area Businesses: If your business operates within a service area rather than a fixed location, utilize the service area management feature in GBP. Define your service area precisely to ensure your profile appears in relevant searches for customers within your coverage zone.

A Roadmap to Success

By leveraging the full spectrum of features offered by Google Business Profile, you transform your online presence from a static listing to a dynamic and engaging digital storefront. Utilize high-quality visuals, communicate effectively with your audience, and harness valuable insights to inform strategic decision-making. Remember, consistently maintaining and optimizing your GBP is an ongoing process. By dedicating time and effort to mastering these advanced features, you position your business for long-term success in the ever-evolving digital landscape.

Chapter 7

Strategies for Engaging with Your Audience in the Digital Age

In today's digitally driven world, customer engagement is no longer a luxury; it's a necessity. Consumers crave connections with the brands they interact with. Building an engaged audience fosters brand loyalty, fuels positive word-of-mouth marketing, and ultimately drives business growth. This guide explores various strategies for effectively engaging with your audience across online platforms, equipping you to foster meaningful conversations and build lasting relationships with your customers.

Understanding Your Audience: Tailoring Your Approach

Effective audience engagement starts with understanding who you're communicating with. Conduct market research to identify your target audience's demographics, interests, preferred communication channels, and online behavior. Here are some additional tips:

» Social Media Listening: Utilize social listening tools to monitor online conversations relevant to your brand and industry. Track brand mentions, identify customer needs, and understand the language your audience uses.

» Customer Surveys and Feedback Forms: Actively solicit feedback from your customers through surveys and feedback forms. This provides valuable insights into their preferences and expectations.

» Analyze Website Traffic: Website analytics tools can reveal valuable information about how users interact with your website. Understand which content resonates with your audience and tailor your engagement strategies accordingly.

Creating Compelling Content: Fueling Conversation Starters

High-quality, engaging content forms the foundation of successful audience interaction. Here are some tips for crafting content that sparks conversation:

» Variety is Key: Don't limit yourself to one format. Experiment with blog posts, infographics, videos, polls, quizzes, and even live streams to cater to different learning styles and preferences.

» Focus on Value: Content should provide value to your audience. Educate, entertain, or inspire them with informative blog posts, industry insights, or behind-the-scenes glimpses into your company culture.

» Pose Questions: Encourage conversation by incorporating questions into your content. Ask for opinions, spark discussions, and invite your audience to share their thoughts and experiences.

» Humanize Your Brand: Let your brand personality shine through. Inject humor, share relatable stories, and showcase the people behind your business to build emotional connections with your audience.

Choosing the Right Platform: Meeting Your Audience Where They Are

Different platforms cater to different communication styles and target audiences. Identify the platforms your audience frequents most and tailor your engagement strategies accordingly. Here are some popular options:

» Social Media: Social media platforms like Facebook, Twitter, and Instagram offer excellent opportunities for real-time interaction and community building. Share engaging content, respond to comments and messages promptly, and participate in relevant conversations.

» Email Marketing: Email marketing allows for more personalized communication. Segment your audience based on interests and tailor your email content accordingly. Run targeted campaigns, promote exclusive offers, and invite subscribers to participate in surveys or polls.

» Customer Review Platforms: Actively monitor and respond to reviews on platforms like Google My Business, Yelp, and industry-specific review sites. Thank customers for positive feedback and address concerns raised in negative reviews professionally.

Responding Promptly and Professionally: Building Trust Through Communication

Prompt and professional responses to inquiries and feedback demonstrate that you value your audience's time and opinions. Here are some strategies for effective communication:

» Respond Timely: Aim to respond to all customer inquiries and feedback within 24 hours, especially on time-sensitive matters.

» Personalize Your Responses: Avoid generic, automated responses. Acknowledge the customer by name, address their specific concerns, and personalize your response to show you've heard their message.

» Maintain a Positive Tone: Even if a customer expresses frustration, maintain a professional and courteous tone throughout your communication. Focus on resolving the issue and demonstrating a commitment to customer satisfaction.

» Multi-Channel Support: Offer multiple channels for customers to reach you, such as social media messages, email, phone calls, or a live chat option on your website.

Strategies for Encouraging Ongoing Engagement

Engagement shouldn't be a one-time event; strive to foster a community around your brand. Here are some tips for encouraging ongoing interaction:

» Run Contests and Giveaways: Host contests and giveaways relevant to your target audience. This incentivizes participation and increases brand awareness.

» Host Live Events: Consider hosting live Q&A sessions, product demos, or industry discussions on social media platforms. Live events foster a sense of connection and allow for real-time interaction.

» Reward Loyal Customers: Implement a loyalty program to reward repeat customers and encourage them to remain engaged with your brand.

» User-Generated Content: Encourage user-generated content by running hashtag campaigns or contests that incentivize customers to share their experiences with your brand.

Chapter 8

Captivate and Convert: Creating Compelling Visual Content for Your Google Business Listing

In the ever-evolving digital landscape, first impressions are paramount. For local businesses, Google Business Listings (GBLs) serve as the digital storefront, often acting as the first touchpoint for potential customers. Here, visual content reigns supreme. High-quality photos and videos showcased on your GBP can significantly impact customer engagement, driving clicks, calls, and ultimately, conversions. This guide explores the importance of visual content in Google Business Listings, offering actionable tips for crafting captivating visuals that convert viewers into loyal customers.

The Power of Visuals in Google Business Listings

Visual content plays a crucial role in attracting and engaging potential customers browsing Google Maps and local search results. Here's how:

» Enhanced Visibility: Listings with high-quality photos are more likely to appear prominently in search results.

Google prioritizes listings with visuals, increasing your discoverability for local searches.

- Grabbing Attention: Eye-catching visuals act as a magnet, drawing potential customers to your listing amidst the sea of competing businesses. High-quality photos and videos instantly capture attention and entice users to learn more about what you offer.
- Building Trust and Credibility: Professional visuals convey a sense of legitimacy and professionalism. Clean, well-lit photos and engaging videos showcase your business in its best light, fostering trust and encouraging potential customers to engage with your listing.
- Telling Your Story: A picture is worth a thousand words. Visuals allow you to showcase your brand personality, product features, and the unique experience your business offers.
- Driving Engagement: Compelling visuals encourage clicks. High-quality photos and videos entice users to visit your website, request directions, or contact your business directly through your GBP.

Crafting Captivating Photos for Your Google Business Listing

- High Quality is Essential: Invest in high-resolution photos that are clear, crisp, and visually appealing. Avoid blurry or pixelated images that detract from the overall professionalism of your listing.
- Showcase Your Business: Feature a variety of photos that showcase your storefront, interior, and the overall ambiance of your business.

» Highlight Your Products or Services: Showcase your products in a visually appealing way. For service-based businesses, consider featuring photos of your team in action or satisfied customers enjoying your services.

» Professionalism Matters: Maintain a consistent aesthetic across your photos. Utilize a well-lit environment and consider using a tripod for stable shots.

» Capture Action and People: Don't be afraid to include photos featuring employees or satisfied customers interacting with your products or services. This adds authenticity and showcases the human element of your business.

Optimizing Photos for Google Business Listings

» Descriptive and Relevant File Names: Use descriptive file names that include relevant keywords associated with your business. This can improve search ranking within Google Maps. For example, instead of "IMG_001.jpg," use "bakery-fresh-bread.jpg" for a bakery listing.

» Alt Text for Accessibility: Add alt text to each photo describing the image content. This improves accessibility for visually impaired users and can also be factored into search engine ranking.

The Power of Video in Google Business Listings

Videos offer a dynamic way to showcase your business and engage potential customers. Here's how to leverage video effectively on your Google Business Listing:

- Showcase Your Offerings: Create short, engaging videos highlighting your products, services, or the unique experience you provide.
- Offer Virtual Tours: Provide a virtual tour of your business, allowing potential customers to explore your storefront or workspace remotely.
- Testimonials and Customer Stories: Feature video testimonials from satisfied customers to build trust and social proof.
- Promote Events and Specials: Announce upcoming events, promotions, or special offers through short video clips.

Optimizing Videos for Google Business Listings

- Keep it Short and Engaging: Aim for videos under 30 seconds to capture attention and maximize engagement within the GBP platform.
- High-Quality Production: Maintain a professional quality in your video production. Ensure good lighting, clear sound, and engaging editing.
- Thumbnails Matter: Choose an eye-catching thumbnail image that accurately represents your video content and entices viewers to click play.
- Captions and Descriptions: Include captions and descriptions for your videos to improve accessibility and searchability within Google Maps.
- Beyond Photos and Videos: Additional Visual

Considerations

- 360° Photos: Consider incorporating 360° photos to offer a truly immersive virtual tour experience.
- Logos and Branding: Ensure your business logo is prominently displayed on your GBP listing. Consistency in visuals reinforces brand recognition.

Chapter 9

Mastering the Mobile Moment: Optimizing Your Google Business Profile for Mobile Users

In today's mobile-first world, the way users interact with local businesses has fundamentally shifted. Gone are the days of desktop searches; smartphones now reign supreme as the primary tool for local business discovery. This necessitates optimizing your Google Business Profile (GBP) for mobile users, ensuring a seamless experience that fosters engagement and drives conversions. This guide delves into essential mobile optimization strategies, empowering you to transform your GBP into a mobile-friendly powerhouse that attracts and retains local customers.

Understanding the Mobile Landscape: The Power of Local Searches

» Mobile Dominance: Statistics reveal a significant shift towards mobile search. Over half of all local searches originate from mobile devices. Optimizing your GBP for mobile is no longer optional; it's a necessity for reaching your target audience.

» On-the-Go Needs: Mobile users typically conduct local searches with immediate needs in mind. They might be searching for "restaurants near me" or "hairstylists open now." Your GBP needs to cater to this immediacy and provide users with the information they need quickly and efficiently.

» Limited Screen Space: Mobile devices offer limited screen real estate. Complicated layouts, cluttered information, or slow loading times will frustrate users and lead to them abandoning your listing.

Optimizing Your Google Business Profile for Mobile Users:

» Accurate and Complete Information: Ensure your NAP (Name, Address, and Phone Number) information is accurate and consistent across all platforms, particularly your GBP. Inaccurate information can lead frustrated users to abandon your listing and seek out competitors.

» Mobile-Friendly Website: Your website should be optimized for mobile viewing. This means ensuring a responsive design that adapts to different screen sizes and offers a smooth user experience on smartphones and tablets.

» Call to Action Buttons: Make it easy for mobile users to take action. Include prominent call-to-action buttons that allow them to call your business, get directions, or book an appointment directly through your GBP.

» High-Quality Photos and Videos: Visuals are crucial for mobile users. Showcase high-quality photos and short, engaging videos that highlight your business offerings. Optimize image size for mobile loading speeds and consider using a 360° photo to offer an immersive virtual tour experience.

Ensuring a Seamless User Experience on Mobile

» Fast Loading Times: Mobile users are impatient. Ensure your GBP listing and website load quickly on mobile devices. Consider compressing image sizes and optimizing website code for speed.

» Click-to-Call Functionality: Enable click-to-call functionality on your mobile GBP listing. This allows users to initiate a call with your business with a single tap, streamlining the communication process.

» Responding to Reviews Promptly: Respond to all reviews, both positive and negative, promptly. This demonstrates responsiveness and builds trust with potential customers, especially those browsing your listing on mobile devices.

» Messaging Functionality: Consider enabling the messaging functionality on your GBP. This allows mobile users to send you direct messages through your profile, offering a convenient way for them to ask questions or schedule appointments.

Leveraging Mobile-Specific Features

» Google Posts: Regularly utilize Google Posts to share updates, events, promotions, and new offerings specifically targeted towards mobile users. Posts with engaging visuals and concise information resonate well on mobile devices.

» Special Offers: Run targeted promotions and special offers directly through your GBP, specifically catering to mobile users searching for deals on the go.

» Service Area Businesses: If your business operates within a service area rather than a fixed location, ensure your service area is accurately defined on your mobile GBP profile. This increases discoverability for relevant local searches on mobile devices.

Tracking and Analyzing Mobile Performance

» Mobile Traffic Analytics: Utilize website analytics tools to track mobile traffic to your website. Analyze user behavior, identify areas for improvement on your website's mobile functionality, and refine your mobile optimization strategy based on data-driven insights.

» GBP Insights: Google Business Profile insights provide valuable data on user interaction with your profile. Track metrics like profile views, website clicks, and call metrics specifically from mobile devices. This helps you understand how mobile users engage with your listing and identify areas for improvement.

The Mobile-First Future

By prioritizing mobile optimization, you ensure your Google Business Profile is discoverable and user-friendly for the ever-growing segment of mobile searchers. Remember, mobile users are looking for immediate answers and seamless experiences. A mobile-optimized GBP equipped with high-quality visuals, clear information, and convenient features positions your business for success in the mobile-first future, attracting and converting local customers on the go.

Chapter 10

Monitoring and Measuring Success for Your Google Business Profile

In the dynamic world of local SEO, success for your Google Business Profile (GBP) hinges on understanding its performance. Just like a captain relies on their compass to navigate, you need key metrics and data analysis to optimize your GBP and achieve your local business goals. This guide equips you with the knowledge and tools to effectively monitor and measure your GBP's success, empowering you to adjust strategies based on data-driven insights for continuous improvement.

Defining Success for Your Google Business Profile

Before diving into metrics, clearly define your goals for your GBP. Are you aiming to increase website traffic, drive more phone calls, or boost brand awareness? Understanding your objectives guides your selection of relevant metrics to track.

Essential Metrics for GBP Performance Monitoring

» Profile Views: Track how often your GBP listing appears in search results. An increase in profile views indicates growing visibility within local searches.

- Website Clicks: Analyze how many users click through from your listing to your website. This metric helps gauge the effectiveness of your listing in driving traffic to your online presence.
- Direction Requests: Monitor how many users request directions to your physical location through your GBP. This can indicate strong local search visibility and potential customer intent to visit your business.
- Call Metrics: Track the number of calls initiated directly from your GBP listing. This provides valuable insight into phone-based lead generation and customer engagement.
- Engagement Metrics: Analyze how users interact with your GBP beyond basic actions. Track metrics like photo views, post engagement (likes and shares), and the number of questions asked through the Q&A section. These metrics offer valuable insights into user interest and content effectiveness.

Understanding User Discovery Sources

- Discovery Source: This metric reveals how users find your GBP listing. Common sources include Google Search, Google Maps, and discovery via direct link. Understanding discovery sources allows you to tailor your optimization strategies for maximum visibility across different platforms.
- Search Queries: See which search terms users employed to find your listing. This data can help you refine your GBP description and optimize keyword usage to improve searchability for relevant local searches.

Leveraging Customer Reviews and Ratings

- » Review Volume: Monitor the number of reviews your GBP receives. A steady stream of positive reviews builds trust and credibility with potential customers.
- » Review Sentiment: Analyze the overall sentiment of customer reviews. Identify areas where you excel and address any recurring negative feedback to enhance customer satisfaction.
- » Review Responses: Track how promptly you respond to customer reviews. Responding to both positive and negative reviews demonstrates responsiveness and builds trust with potential customers.

Utilizing Free and Paid Analytics Tools

- » Google My Business Insights: GBP offers a suite of built-in analytics tools that provide valuable data on user interaction with your listing. Utilize these insights to understand your audience, measure the effectiveness of your content, and identify areas for improvement.
- » Google Search Console: This free tool from Google offers valuable insights on organic search performance. Analyze search queries and optimize your website content and GBP description to improve ranking for relevant local keywords.
- » Paid Analytics Tools: Consider investing in paid analytics tools for deeper insights. Services like Google Analytics offer detailed user behavior analysis and allow you to track conversion rates and attribute revenue generated from your GBP listing.

Turning Data into Action: Optimizing Strategies for Growth

Data analysis empowers you to refine your GBP optimization strategies. Here's how:

- Content Optimization: Analyze popular search queries and user engagement with your GBP content (posts, photos, etc.). Focus on creating content that resonates with your audience and addresses their local search needs.
- Keyword Targeting: Based on search query data from Google Search Console and GBP insights, refine your GBP description and website content to target relevant local keywords.
- Addressing Customer Needs: Analyze customer reviews and identify recurring themes. Use these insights to improve your products, services, or address common customer pain points.
- Aligning GBP Strategy with Local SEO: Ensure your GBP listing information is consistent with your website and other online directories. This strengthens local search ranking signals and improves overall online visibility.

Conclusion: A Culture of Continuous Improvement

Monitoring and measuring your Google Business Profile performance is not a one-time endeavor; it's an ongoing process. By regularly analyzing key metrics and adjusting your strategies based on data-driven insights, you can continuously optimize your GBP, improve user experience, and ultimately achieve your local business goals. Remember, a data-driven approach equips you to navigate the ever-evolving landscape of local SEO and positions your business for sustained success in the digital marketplace.

Chapter 11

Amplifying Your Local Presence: Expanding Reach with Google Ads for Your Google Business Profile

In today's hyper-competitive landscape, simply claiming your Google Business Profile (GBP) isn't enough to ensure local business success. To truly stand out and attract new customers, consider leveraging the power of Google Ads. This guide introduces Google Ads for local businesses, equipping you with the knowledge and strategies to run effective ad campaigns that amplify your GBP's reach and drive targeted local traffic.

Understanding Google Ads for Local Businesses

Google Ads offers a robust advertising platform that allows businesses to promote themselves on Google Search results pages and Google Maps. Here's what makes it particularly beneficial for local businesses:

» Targeted Reach: Reach potential customers actively searching for products or services like yours within your specific local area.

- Enhanced Visibility: Increase the prominence of your GBP listing in search results, attracting more local customers browsing Google Maps and conducting local searches.
- Measurable Results: Track the performance of your ad campaigns, analyze key metrics, and optimize your strategy for maximum return on investment (ROI).
- Flexible Budget: Set a budget that aligns with your advertising goals and adjust it as needed, ensuring you stay in control of your advertising spend.

Campaign Types for Local Business Promotion

Search Network Campaigns: Target users actively searching for keywords related to your business and location. Craft compelling ad copy that highlights your unique selling points and encourages users to click through to your GBP listing or website.

- Local Services Ads: This specialized ad format is specifically designed for service-based businesses. Showcase your expertise, qualifications, and customer reviews directly within search results, attracting high-quality leads actively seeking local services.
- Display Network Campaigns: Reach potential customers across a vast network of websites and apps relevant to your local market. Target users based on demographics, interests, and online behavior to build brand awareness and drive them to your GBP listing.

Strategies for Effective Local Business Ad Campaigns

» Location Targeting: Ensure your ads are only displayed to users searching within your designated service area. This ensures your budget is spent on reaching potential customers who can realistically patronize your business.

» Keyword Research: Conduct thorough keyword research to identify relevant keywords with high local search volume. Target these keywords within your ad copy to increase the chances of your ads appearing for searches related to your products or services.

» Compelling Ad Copy: Craft clear, concise, and engaging ad copy that highlights your unique offerings and value proposition. Include a strong call to action that entices users to click through to your GBP listing or website.

» Call Extensions: Utilize call extensions to allow users to instantly initiate a phone call to your business directly from your ad. This simplifies the conversion process and caters to users seeking immediate service.

» Location Extensions: Include location extensions in your ads to display your business address alongside your listing. This allows users to easily see your physical location and potentially visit your business directly.

» Reviews Extensions: Showcase positive customer reviews from your GBP within your ad copy. This builds trust and credibility, encouraging users to choose your business over competitors.

» Mobile Optimization: Ensure your ad campaigns are optimized for mobile devices. A significant portion of local searches originate from smartphones, so a mobile-friendly ad experience is crucial for capturing potential customers on the go.

Leveraging Your Google Business Profile with Google Ads

- Strong Listing Foundation: Before launching your ad campaign, ensure your GBP is well-optimized with accurate information, high-quality photos, and engaging content. A strong GBP listing provides a seamless landing page experience for users clicking on your ads.
- Call to Action Consistency: Maintain consistent call-to-actions across your ads and GBP listing. This ensures a smooth user experience and encourages conversions, whether it's prompting users to visit your website, call your business, or book an appointment directly through your GBP.
- Track and Analyze Performance: Monitor the performance of your ad campaigns closely. Analyze key metrics like click-through rates, conversion rates, and cost-per-acquisition (CPA). Utilize these insights to optimize your campaigns for better targeting and improved ROI.

The Synergy of Google Business Profile and Google Ads

By strategically combining the power of Google Business Profile optimization with targeted advertising through Google Ads, you create a robust local marketing strategy that amplifies your reach, attracts new customers, and fuels local business growth. Remember, consistent monitoring, data-driven optimization, and a commitment to providing a seamless user experience across your GBP and ad landing pages are key to unlocking the full potential of this powerful marketing combination.

Chapter 12

Charting the Course: Advanced Tactics and Future Trends in Google Business Profile Optimization

In the ever-evolving realm of local SEO, staying ahead of the curve is paramount for businesses vying for local prominence. This guide delves into advanced tactics for optimizing your Google Business Profile (GBP) alongside predictions for the future of local search and Google Business Listings. By embracing cutting-edge techniques and preparing for upcoming trends, you can future-proof your local SEO strategy and position your business for sustained success.

Advanced GBP Optimization Tactics for Savvy Businesses

» Structured Data Implementation: Utilize schema markup, a form of structured data, to provide search engines with richer information about your business. This can enhance your GBP's visibility in search results and improve user understanding of your offerings.

» Backlink Building: While backlinks hold less weight than previously, acquiring high-quality backlinks from local websites and industry publications can still provide a slight SEO boost and enhance your GBP's overall credibility.

» Q&A Optimization: Proactively populate your GBP's Q&A section with frequently asked questions and informative answers. This demonstrates expertise, addresses potential customer concerns, and improves organic ranking within local searches.

» Image Recognition and Optimization: Leverage Google Lens' image recognition capabilities. Optimize GBP photos with relevant keywords and alt text descriptions to ensure your images appear in relevant image searches, potentially driving more traffic to your listing.

» Responding with Videos: Consider incorporating short video responses to customer reviews on your GBP. This personalized approach can enhance customer engagement, demonstrate responsiveness, and showcase your brand personality.

» Community Engagement: Actively participate in local online communities and forums relevant to your industry. Offer valuable insights, share informative content, and subtly promote your business while building brand awareness and establishing yourself as a local expert.

» Highlighting Special Attributes: Go beyond basic business information. Utilize the special attributes section of your GBP to showcase unique features and amenities offered by your business. This could include wheelchair accessibility, Wi-Fi availability, kid-friendliness, or eco-friendly practices.

» Promotions and Offers: Run targeted promotions and special offers directly through your GBP. Announce limited-time deals, discounts, or loyalty programs to entice new customers and encourage repeat business.

Envisioning the Future of Local Search: Emerging Trends and GBP Evolution

» The Rise of Voice Search: Voice search is rapidly gaining traction. Optimize your GBP listing with natural language search queries in mind. Focus on conversational keywords and ensure your business information is readily available for voice search assistants.

» Visual Search and Local Discovery: With advancements in image recognition technology, visual search is poised to become a significant factor in local discovery. Maintain high-quality, well-optimized photos and virtual tours within your GBP to capture attention and attract potential customers searching visually.

» The Focus on User Generated Content (UGC): User-generated content, including customer reviews, photos, and videos, is likely to hold increasing value in future local search algorithms. Actively encourage customer engagement to foster a steady stream of positive UGC that builds trust and enhances your GBP's credibility.

» Integration with Appointment Booking and Online Ordering: Expect seamless integration between Google Business Listings and online appointment booking or ordering systems. This will allow users to schedule appointments or place orders directly through your GBP, streamlining the customer journey and fostering conversions.

» Personalized Search and GBP Recommendations: As Google tailors search results to individual users, anticipate GBP recommendations personalized based on search history and location. Optimizing your GBP for diverse customer segments can help ensure your listing appears in relevant searches for a wider audience.

» AI-Powered Chatbots and Customer Service: The future of local business engagement might involve AI-powered chatbots integrated into GBP listings. Explore implementing chatbots to answer basic customer inquiries and streamline communication, particularly during off-business hours.

Embracing Innovation and Staying Ahead

The local SEO landscape is constantly evolving. By embracing advanced optimization tactics and preparing for upcoming trends, you demonstrate a commitment to remaining competitive and delivering an exceptional customer experience. Remember, a future-proof local SEO strategy goes beyond simply claiming your GBP. It requires continuous learning, adaptation, and leveraging innovative techniques to ensure your business thrives in the ever-changing world of local search and Google Business Listings.

Chapter 13

Free Resources

Here is a video tutorial created by Erica Glessing on how to create a strong business profile on Google:

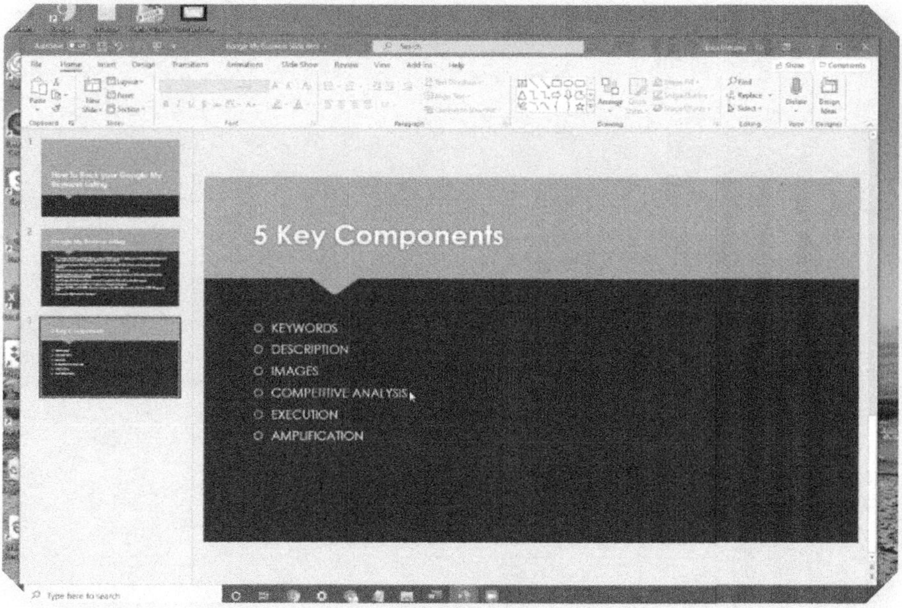

https://www.youtube.com/watch?v=FJPSUb2Ed7Q

For a complimentary consultation on improvements to your business online presence, you can reach Erica here: https://SEOforLeadGen.com

Made in the USA
Monee, IL
04 November 2026

67381358R00036